DOMINATOR
The Story of the Consolidated
B-32 BOMBER

This fine study of B-32-25-CF 42-108547 shows the Briggs-built A-13A lower ball turret in its fully extended position. The plane served in the test program at Eglin Field, Florida, becoming aircraft No. 591 of Squadron "E," 611th AAFBU, AAFPGC.

General Dynamics via Mike Moore

Eight TB-32s stand on the factory ramp at Fort Worth in February 1945, awaiting final checks and delivery to the 2519th AAFBU. Serials of the nearest four aircraft are, nearest to farthest, 42-108500, 498, 493 and 495.

General Dynamics via William T. Larkins

DOMINATOR

The Story of the Consolidated

B-32 BOMBER

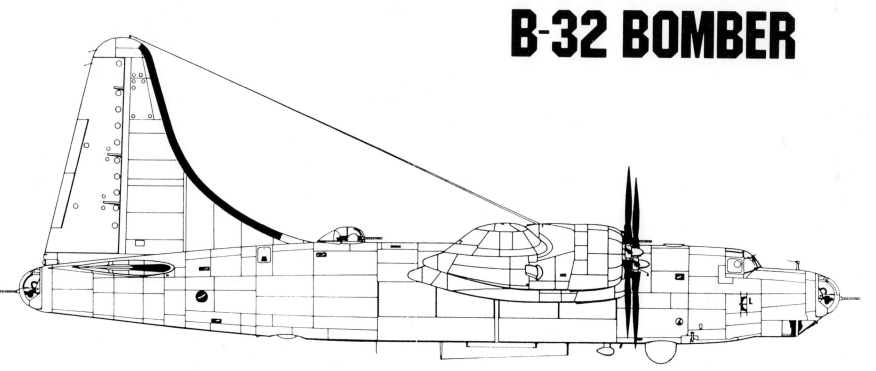

By
Stephen Harding & James Long

Pictorial Histories Publishing Company
Missoula, Montana

LIBRARY OF CONGRESS
CATALOG CARD NUMBER 83-62473

ISBN 0-933126-38-7

First Printing: February 1984
Second Printing: January 1986
Third Printing: October 1989
Fourth Printing: April 1996
Fifth Printing: April 2003
Sixth Printing: August 2005

FRONT COVER: A B-32 is attacked by Japanese fighters over Tokyo while on a
photo-recon mission. Incident took place a few days after the atomic bomb was
dropped on Hiroshima. Artist: William J. Reynolds, Waldorf, Maryland

Typography: Arrow Graphics
Layout: Stan Cohen

Plane profile by Alan Chesley, Annapolis, Maryland
Detailed plane drawings by James I. Long, Ocean Springs, Mississippi

PICTORIAL HISTORIES PUBLISHING CO., INC.
713 South Third St. West, Missoula, Montana 59801
Phone (406) 549-8488   Fax (406) 728-9280
Website: pictorialhistoriespublishing.com
E-Mail phpc@montana.com

Table of Contents

About the Authors

Stephen Harding holds both BA and MA degrees in modern military history from the University of California, and is currently a staff historian in the U.S. Army Office of the Chief of Military History. His articles on various topics of naval and military history have appeared in a wide range of professional and popular journals, and he is the author of *Gray Ghost: The RMS Queen Mary at War.* Harding lives in Washington, D.C., with his wife, Mary, and daughter, Sarah, and is currently at work on a history of naval airships.

James Long, owner of AIR'TELL Publications and Research Service, has confined his recent work to studies and writings about WWII Japanese aircraft. He has contributed to a number of new books and restoration projects, most notable of which have been a chapter on serial numbers in a book on the Zero Fighter by Robert C. Mikesh and the re-creation of airframe markings for the recent restoration of the National Air and Space Museum's NIK2-J naval fighter. Long, a native of Stillwater, Okla., and his wife, Margie, make their home in Ocean Springs, Miss. They have two children, Karin and Michael.

Acknowledgments

The authors wish to thank the following individuals and organizations, without whose help this work would have been far less complete: Dana Bell; Roy G. Cherry; Walter D. France; Robert A. Garfield; James W. Hill; William T. Larkins; Richard Schulenberg; Leonard E. Williams; Earl R. Meisenheimer; William T. Y'Blood; SSgt John Cooley, 832nd Air Division, George AFB, California; Rob Mack; Joe Thornton and Jack Isabel of General Dynamics; Judy G. Endicott and the staff of the Albert F. Simpson Historical Research Center, Maxwell AFB, Alabama; Paul M. Stickel and the 312th Bombardment Group Association; Bruce Reynolds and the staff of the San Diego Aerospace Museum; Marilyn Phipps of Boeing Aircraft, and the U.S. Branch of the International Plastic Modelers' Society for permission to reprint nose art drawings.

Foreword

With this 4th printing, we gratefully acknowledge all of the people who have made this book a success by buying copies over the years since its first publication. And we are especially thankful for those who have written to offer additional technical data or their prized photos and personal recollections of the long-departed B-32. These post-publication contributors and promoters are Lyman P. Combs, Anthony R. Conti, 2nd Lt Edwin L. DuFeu, Lt Col David W. Ecoff, Capt Joe E. Elliott, Capt Robert L. Horner, David W. Menard, Douglas D. Olson, Charles H. Paul, SSgt Bob Plummer, Cpl Donald R. Touhy, and Lt Col Floyd B. Whitlow. Special commendations go to Robert A. Garfield of Canton, Ohio, a B-32 aerial engineer, and to Mike Moore of Fort Worth, Texas, a B-32 enthusiast. Both men have been most liberal contributors to this project—before publication and after.

James I. Long, Ocean Springs, Miss

Three TB-32s on a formation flight over the Dallas-Fort Worth area in March 1945. The aircraft are 42-108520, 515 and, in the foreground, 524. 515 and 520 are TB-32-10-CF, while 524, the last of the forty trainers built, is a TB-32-15-CF. Note that it has deicer boots on the tail members, a trait characteristic of the last four TB-32s. All earlier Dominators, including the first fourteen bombers, were manufactured without them.

General Dynamics

CHAPTER ONE

Origin of the B-32

In any discussion of American heavy bombers of the Second World War the aircraft most frequently mentioned are the B-17 Flying Fortress, the B-24 Liberator and the B-29 Superfortress. This is understandable, for those aerial workhorses were produced in great numbers, served in nearly every theater of war, and formed the backbone of the U.S. Army Air Force's strategic bombing offensive against the Axis powers. It is equally understandable why the Consolidated B-32 Dominator is the least remembered participant in that great offensive, for the 115 production B-32s delivered to the Army Air Forces amounted to only slightly more than two-tenths of one percent of the 50,750 bombardment aircraft accepted from contractors during the Second World War.

Indeed, at first glance it seems surprising that the Dominator played even the small role it did in the history of World War II aerial operations. Plagued from inception by a series of technical problems, the B-32 was the focal point of heated debate throughout its operational life. Though originally intended to join the B-29 in the Very Heavy Bomber (VHB) role, the Dominator was ultimately made superfluous by her running mate's operational success. It saw only limited combat in the last days of the war against Japan and, when that war ended, it was discarded with almost embarrassing swiftness.

Yet a closer examination of the B-32's history reveals that the Dominator does indeed have a place in the story of combat aviation, even if it must stand in the shadow cast by other, more famous aircraft. The record shows that the B-32 was not the total failure her detractors claimed. When pressed into service by the Far East Air Forces—simply because it was available—the Dominator proved itself a rugged and stable, though admittedly tempermental, bombing platform. And, though its combat record in those closing days of the war is generally unspectacular, the B-32 is distinguished for having fought, and won, the final aerial combat against the Japanese Empire.

The beginning of the Dominator's chequered career can be traced to the early 1930s, by which time it had become all too clear to Air Corps planners that victory in future wars would depend largely on America's ability to project its airpower over vast distances. In the minds of those planners, the projection of airpower specifically meant the projection of *strategic bombardment* airpower. Thus, throughout the decade, emphasis was placed on the development of large, long-range bombers capable of attacking distant targets with a significant bomb load. The Boeing B-17 and the Consolidated B-24 and, to a greater extent, the Boeing B-15 and the Douglas B-19, were all important steps toward what the Air Corps called the "Very Long Range" (VLR) or, occasionally, the "Very Heavy" bomber.

Yet by 1939 no suitable VLR bomber had reached the production stage. The B-15 and B-19 had gone the way of all pioneering test

The 1/35th scale model of the XB-32 that underwent wind tunnel tests in December 1940 bore little resemblance to production Dominators. The model, which was made of birch and finished with lacquer, had a wing span of 3.8 feet. The landing gear and gun turrets were removable, to simulate their retraction, and the stabilizers, rudders and elevators were adjustable.

USAF/Albert F. Simpson Historical Research Center

Partially completed revised full-scale mock-up of B-32, Fort Worth Army Airfield, April 1945.

USAF/AFSHRC

Revised full-scale B-32 flight deck mock-up at Fort Worth AAFld, April 1945.

USAF/AFSHRC

Cockpit of full-scale flight deck mock-up.

USAF/AFSHRC

beds and the B-17 and B-24, though excellent aircraft, simply were not VLR material. The Air Corps needed long range airplanes capable of bombing from both high (20,000 to 30,000 feet) and medium (5,000 to 7,000 feet) altitudes, and which could protect themselves during the journey. The outbreak of war in Europe in September 1939 provided just the shock needed to transform the VLR project from theory into practice, however, and on 10 November General H.H. Arnold, the Air Corps Chief of Staff, started the ball rolling. General Arnold asked the War Department for permission to initiate the development of a four-engine, long-range heavy bomber meant to surpass in all respects the performance of then-current models of the B-17 and the B-24. The War Department granted the authority on 2 December, and the "Request for Data R-40B" was circulated among the nation's leading aircraft companies on 29 January 1940.

In keeping with the tenets of the VLR philosophy, the Army's specifications for the new bomber stressed high speed and long range rather than massive bomb load. The aircraft, according to the Army, would have to be able to carry 2,000 pounds of bombs over a range of 5,333 miles at a speed of 400 miles per hour. The new bomber would have to offer complete mechanical reliability and, in keeping with the Air Corps' adherence to the principle of high altitude strategic bombardment, would have to be pressurized. On 8 April 1940 preliminary design studies were submitted by Boeing, Lockheed, Douglas and Consolidated. On 27 June an Army Air Corps Evaluation Board designated the proposed aircraft in order of preference as, respectively, the XB-29, XB-30, XB-31 and XB-32, and signed preliminary engineering data contracts with each firm. Both Lockheed and Douglas subsequently withdrew from the competition, leaving the Boeing and Consolidated proposals as the Army's only viable alternatives.

The aircraft proposed by Consolidated was known within the company as the Model 33, and was at first glance an obvious descendent of the B-24. The shoulder-mounted, high-lift/low-drag Davis wing and twin endplate fin and rudder assemblies were the most visible similarities. And, like the B-24, the B-32 was to have tricycle landing gear and dual "roll-up" bomb bay doors. Yet the new bomber could not accurately be called a "souped-up" Liberator, for the differences between the two aircraft far outweighed the similarities. The XB-32 was to have a cylindrical fuselage 83 feet in length, compared to the B-24's 66 foot box-shape and, as originally planned, a rounded nose instead of the

Liberator's stepped windscreen. At 135 feet the XB-32's wingspan would be 25 feet wider than that of the B-24, and its four Wright B-670 turbo-supercharged Duplex Cyclones would produce 1,000 more horsepower per engine than the 1,200 hp Pratt and Whitney R-1830s of the Liberator. The new bomber was to have remotely operated retractable gun turrets in place of the B-24's manned positions, and its crew spaces were to be pressurized for high-altitude bombing. All these features were of course reflected in the XB-32's projected gross weight, which was expected to be some 45,000 pounds higher than the Liberator's 56,000 pounds.

Consolidated's design proposal was approved by the Air Corps on 6 September 1940, at which time the company was awarded a contract for the production of two prototypes. This was later changed to three aircraft, with the first XB-32 to be delivered within 18 months of the contract date, the second within 90 days of the first, and the third within 90 days of the second. The contractor was also required to produce a 1/35th scale wooden model of the bomber for wind tunnel testing, which it did, and the tests were carried out by the Air Corps Materiel Division at Wright Field during the last ten days of December. The tests revealed a problem that would plague the twin-tailed XB-32s, for though the proposed bomber's longitudinal control was rated satisfactory, its directional stability was judged to be "insufficient."

The mock-ups built by Consolidated in late December incorporated many minor changes suggested by the Wright Field report but retained the twin fin and rudder assemblies. On 6 January 1941 an Army inspection team examined and approved the revised mock-ups at Consolidated's San Diego plant despite misgivings about the twin-tailed arrangement. Powerplant mock-ups were likewise inspected and approved on 17 April, and by June the Army felt confident enough about the new aircraft to order that 13 YB-32s be developed parallel to the three XB-32 prototypes.

The first XB-32 rolled off Consolidated's San Diego assembly line on 1 September 1942, almost exactly six months behind schedule. The delay, brought about by problems with the aircraft's pressurization and turret retraction systems, prompted the Army to request that flight tests begin "as soon as possible, even if the aircraft must be flown in a 'stripped' condition." The Army's attitude was understandable, since the still untried B-32 had already assumed a major role in the Air Corps' long-range contingency planning for any eventual participation in the war in Europe. The plan, completed in August 1941 and given the code number

AWPD/1, was based on a program of precise strategic bombing aimed at Nazi Germany's industrial heartland and required a total of 6,834 bombers organized in 98 groups. Ten of these were to be "Medium" groups equipped with B-25s and B-26s, twenty would be "Heavy" groups of B-17s and B-24s, and the remainder "Very Heavy" groups built around B-29s and B-32s. Since the XB-32 was the first of the two "very heavies" to be completed, the XB-29 having suffered its own delays, it was imperative that the Consolidated aircraft reach the production stage quickly if AWPD/1 were to proceed on schedule. The XB-32 accordingly began taxi tests on 3 September and, following a few minor adjustments, took to the air for the first time four days later.

Consolidated drafting personnel at work. The many changes and modifications made to the B-32 design kept the contractor's engineering section busy. Production startup was originally scheduled for the fall of 1942, but the almost continual series of Dominator design changes delayed work until February 1943. General Dynamics

CHAPTER TWO
XB-32

The first XB-32 prototype, serial number 41-141, was just over 83 feet long. Contrary to popular belief, the aircraft was built with a stepped windscreen rather than the B-29 style rounded greenhouse nose specified in Consolidated's original design. The cylindrical fuselage had a maximum diameter of 9 feet 6 inches, with flush-riveting on the forward section and standard brazier-head riveting aft of the flight deck. Crew compartments were located fore and aft of the dual tandem bomb bays and connected by a catwalk running along a heavy built-up beam. The bomb bay doors were of the B-24 hydraulically powered "roll-up" type, and were built of aluminum sheet backed by hat-shaped stiffeners. The aircraft's fully cantilevered Davis wing spanned 135 feet and was built up on two spars with metal ribs and flush-riveted stressed aluminum skin. The wing center section was permanently attached to the fuselage and housed four self-sealing fuel tanks for the bomber's four 2,200 hp Wright R-3350-13 radial Cyclone engines. Four Alclad-covered Fowler flaps, two on each side of the fuselage, spanned the center section of the wing from fuselage to ailerons. The XB-32 retained the twin-tails of the original design, and each rudder was fitted with trim tabs. The empty weight of the first "Terminator," as the B-32 was originally named, was 64,960 pounds, gross weight was 101,662 pounds, and maximum permissable weight was 113,500 pounds.

The XB-32 made its first flight on 7 September 1942 from San Diego's Lindbergh Field. All went well until about twenty minutes into the flight, when one of the rudder trim tabs malfunctioned and the aircraft was buffeted by severe tail flutter. The pilot, Consolidated's Russ Rodgers, prudently brought the huge plane in for an emergency landing at North Island Naval Air Station. Testing resumed following alterations to the trim tab system, but the seemingly endless series of minor problems that continued to plague 41-141 prompted the Army to cancel its order for 13 YB-32 service test aircraft in February 1943. Though the YB-32s were ostensibly to be added to the first production contract, it was all too clear that the AAF was becoming increasingly concerned about the future of the B-32 project. That future was even more in doubt following the loss of 41-141 on 10 May 1943. A still unexplained flap malfunction caused the first XB-32 to crash just after take-off, injuring six crewmen and killing Consolidated's Senior Pilot, Richard McMakin.

The loss of 41-141 was to prove a major setback for the B-32 program, for many vital test records were destroyed in the crash and most of the completed tests had to be repeated using the second prototype. But production of the second XB-32 was also behind schedule, and it would be another five weeks before flight testing could be resumed. This delay was to have a far reaching impact on the aircraft's future, for the development of Boeing's XB-29 was proceeding rapidly and some influential Army officers were already

The crash of the first prototype meant that testing had to be completed using the second XB-32, 41-142, seen here on a flight over southern California. When its usefulness was at an end the aircraft was broken down, crated, and delivered to the AAF for storage in San Diego. It was disposed of by reclamation on 21 January 1946. General Dynamics via William T. Larkins

The third XB-32, 41-18336, tries out its new single-unit vertical tail. The last digit of the radio call number, painted on the tail, is obscured by part of the modification. Note that the aircraft retains the greenhouse-style nose and cockpit enclosure, and is fitted with a dummy dorsal turret.
General Dynamics

beginning to advocate the outright cancellation of the B-32 in favor of the B-29.

The second XB-32, serial number 41-142, made its first flight on 2 July 1943 sporting the same type of twin rudder assembly as its predecessor. Like 41-141, 41-142 was pressurized and equipped with remote-controlled retractable gun turrets in the dorsal and ventral positions, supplemented by a manned position in the tail. This defensive armament arrangement was to become another bone of contention between Consolidated and the Army, however, for in November 1943 an Army Air Forces inspection team found the aircraft's turret system to be unacceptable. The team's report stated that the XB-32's nose and tail positions offered "insufficient" firepower, that a "blind spot" existed just forward of the wing tips that none of the turrets could cover, and that none of the bomber's guns had an adequate supply of ammunition. Indeed, the inspection team's report flatly concluded that "the defensive firepower is inadequate for a long range bombardment airplane operating beyond the range of fighter escort" and suggested the remotely controlled turrets be replaced by manned installations.

The first flight of the third XB-32, serial 41-18336, thus took place on 9 November 1943 under an increasingly dark cloud of controversy. Continuing technical problems had delayed delivery of the aircraft and, in the Army's view, it incorporated design features already found to be unsatisfactory. In an effort to save the B-32 project from cancellation Consolidated agreed to yet another AAF engineering inspection of the two existing prototypes. The inspection team's report, which was submitted on 3 December

1943, found the aircraft to be "obsolete when compared with 1943 combat airplane requirements," and recommended a host of major changes. These changes included: 1) a single vertical tail in place of the twin vertical tails; 2) the deletion of any pressurization, current tactical doctrine having made it unnecessary; 3) the replacement of remotely controlled turrets by manned installations; 4) the addition of heated wing deicers; 5) a complete redesign of the engine nacelles; 6) the use of four-bladed propellers; 7) improved fuel and oil systems; 8) an all-electric bomb release system; 9) installation of an M-series bombsight and automatic flight control system; 10) improvement of bombardier's forward and side vision through installation of the Emerson model 128 nose assembly then in use on the B-24; 11) simplification of maintenance by concentrating on accessibility and utilization of standardized replacement parts; and, 12) the improvement of emergency exits.

The changes suggested in the Army's inspection report resulted in the virtual redesign of the B-32 by Consolidated-Vultee. While the second XB-32 prototype continued its flight test program in the twin-tailed configuration, 41-18336 was slated to become the test bed for the required changes. After some 25 flights with twin tails the third XB-32 was taken into the shops and given a single B-29 style 16½-foot vertical tail. This unit proved inadequate, and the aircraft was soon fitted with a Consolidated-designed 19½-foot tail. Other modifications to 41-18336 included the deletion of all pressurization, the relocation of certain interior fixtures, and the experimental installation of various manned gun turrets.

The first XB-32, 41-141, runs up her engines on Consolidated's San Diego ramp soon after rolling off the production line. On 10 May 1943 the aircraft crashed on takeoff, injuring six crewmen and killing Consolidated's Senior Pilot.

General Dynamics via William T. Y'Blood

CHAPTER THREE
Production B-32s

The AAF had placed initial orders for production B-32s in March 1943 and, following the up-grading of the third prototype, these orders were amended to cover aircraft incorporating the recommended modifications. The orders called for production of 300 B-32s, including some TB-32 transition trainers. Orders for additional aircraft soon followed, bringing the total number of Dominators scheduled for delivery in 1943 and 1944 to some 1,200. But these orders were cancelled in October 1945 upon the termination of the B-32 program and, as a result, only 115 production planes were delivered.

Consolidated-Vultee had produced all three XB-32 aircraft at its San Diego plant, but construction of all production articles except 44-90486 was undertaken at the company's facility at Fort Worth, Texas. Aircraft built at Fort Worth bore the manufacturer's code letters -CF, while the three XB-32s and the one production model built at San Diego carried the -CO suffix.

The first production B-32, 42-108471, was originally built with the 16½-foot B-29 style tail, but this was later changed to the standard 19½-foot version. All subsequent production B- and TB-32 aircraft were identical in major structural detail and differed only in internal equipment and minor changes to certain sub-systems. Standard structural details were as follows:

Fuselage: The fuselage was 83 feet 1.015 inches long, circular in cross section, with a maximum diameter of 9 feet 6 inches. The riveting patterns on production aircraft were the same as for the XB-32. The flight deck featured a stepped windscreen, and the greenhouse nose of the XB-32 was replaced by a very large manned ball turret manufactured by the Sperry company. The early pattern greenhouse tail stinger was also replaced by a Sperry manned installation. The tandem bomb bays were covered by hydraulically operated roll-up doors like those used on the B-24. A heavy built-up beam ran the full 297-inch length of the bomb bay and carried a catwalk connecting the forward and aft crew compartments.

Wing: The Dominator's shoulder-mounted low-drag Davis wing spanned 135 feet, was fully cantilevered and internally braced, and was built up on two spars with metal ribs and flush-riveted stressed skin. The wing center section was permanently attached to the fuselage, though the tips and both outer panels were removable. The wing had the Consolidated-NACA thermal anti-icing system originally developed for the B-24 and PBY, and total wing area was 1,422 square feet. The B-32 was also fitted with four Alclad-covered Fowler flaps, two on each side of the fuselage spanning the wing center section from the fuselage to the ailerons. The flaps were hydraulically operated and electrically controlled, and each travelled on three tracks. The two outboard and two inboard flaps were mechanically joined to form two sets, and the aft tips of the engine nacelles extended with the flaps. The maximum flap deflection was 40 degrees, and total wing area with the flaps extended was 1,544.94 square feet. The ailerons were of monospar metal construction, fabric covered, and statically and dynamically balanced. Tabs on each aileron provided trim and balance, and each aileron had a maximum up or down deflection of 20 degrees.

Empennage: The tail assembly fitted to all production B-32 aircraft consisted of a single vertical fin and fully cantilevered stabilizers. The rudder and elevators were fabric-covered with Alclad-reinforced noses, statically and dynamically balanced. Each elevator had one combination trim and servo tab, and the rudder on most production models was equipped with one servo and one manual tab.

Engines: Production B-32s mounted four air-cooled Wright R-3350-23A Cyclone radials, each of which had two exhaust-driven turbosuperchargers. The engines were of conventional design with two staggered banks of nine cylinders each, and each developed 2,200 hp on 100 octane fuel. The engines drove four-bladed Curtiss Electric propellers which, at a diameter of 16 feet 8

inches, were the largest used on a production aircraft up to that time. Each prop was fitted with a Curtiss synchronizer which automatically altered the blade angle to provide constant speed control. The two inboard propellers could be reversed to −15.8 degrees for reverse thrust braking, and were another innovation first used on the B-32.

Fuel Tanks: The B-32's four self-sealing fuel tanks consisted of 12 cells and were housed in the center section of the wing. The total rated capacity of the tanks was 5,460 U.S. gallons, though about 90 gallons were trapped in the tanks by the plane's normal 4-degree nose-up flight attitude and could not be used. All production B-32s could be equipped to carry four removable self-sealing bomb bay tanks, each with a capacity of some 750 U.S. gallons. The tanks were shackled to the plane's bomb racks and provided with quick-disconnect couplings so they could be dropped in flight.

Oil Tanks: Production B-32s were provided with four self-sealing oil tanks in the wing center section, one just aft of each engine nacelle, with a combined capacity of 306 U.S. gallons. Each tank

The decision to adopt the tall tail was not made in time to influence the initial configuration of the first production aircraft, 42-108471, seen here with the B-29 style tail.

General Dynamics via William T. Y'Blood

The first production aircraft as it appeared in 1945. Though built as a B-32-1-CF, 42-108471 was upgraded virtually to B-32-20-CF specifications during the course of her service life. These modifications included the installation of radar and electronic navigation aids, aft scanner windows and a second rubber tab. Note, however, that the teardrop-shaped radome is unlike those found on other Dominators. This aircraft was unique in having elongated tips on its horizontal tail surfaces, though they are hardly visible in this shot. The modification apparently proved pointless, for it did not appear on any other Dominator.

General Dynamics

The second production B-32-1-CF, 42-108472, displays the true pattern of the first block of aircraft. As built, they had only one rudder tab, no flight instruments for the co-pilot, and no radar or special navigation aids. Though this aircraft was actually the first production article to be delivered to the AAF, on 19 September 1944, it did not last long. On the day of delivery 42-108472 was destroyed when its nose gear collapsed on landing. This photo shows that 472 was already having problems with its hydraulics, for the tail skid was not supposed to extend independently of the landing gear. General Dynamics

The fourth production B-32, 42-108474, awaits delivery to the AAF at the ATSC facility, Vandalia, Ohio, in December 1944. Note that guns have not yet been installed in her turrets. USAF Museum via William T. Y'Blood

The sixth production B-32-1-CF, 42-108476, as it looked when delivered to the AAF on 21 December 1944. The plane was assigned to the ATSC facility at Vandalia, though it was later lent to the AAFPGC at Eglin Field, Florida, for participation in the B-32 service tests. When the war ended 476 was declared excess and sent to the RFC disposal center at Walnut Ridge, Arkansas. Note that the B-32's belly turret protruded slightly when retracted, a characteristic common to the B-24 as well. USAF

was part of an independent oil system which provided constant lubrication and partial cooling for each engine and its related accessories.

Landing Gear: The tricycle landing gear system of the B-32 consisted of three fully retractable sets of dual wheels and a retractable tail skid. The main gear consisted of two sets of dual 56 inch tires and tubes and two sets of dual wheel and brake assemblies, all of which were interchangable with those of the B-24. The nose gear swiveled through a full 360 degrees, and consisted of two 39 inch tires and tubes on co-rotating wheels that were splined to the axle to prevent shimmying. The tail skid was fitted with an air-oil shock strut, and retracted with the main gear.

Armament: The standard defensive armament of production combat-equipped aircraft consisted of ten .50 caliber Browning M-2 machine guns mounted two each in five locally controlled turrets. The nose and tail units were electric-hydraulic A-17 or A-17A turrets built by Sperry, while the lower ball was a Briggs-built fully retractable A-13 or A-13A. The upper turrets were electrically operated Martin A-3D or A-3F units with streamlining "teardrops." TB-32s carried some 700 pounds of ballast to compensate for the lack of turrets.

Bombing Equipment: The standard bombing equipment on production B-32 aircraft consisted of a Norden M-9 bombsight on a B-7 stabilizer mount, positioned in the bombardier's compartment just below and aft of the nose turret, and an A-4 all-electric release system. Bombing instruments were above and to the left of the bombsight, and an entrance/escape hatch occupied the left rear of the compartment. The bomb bay was divided into quadrants by the central catwalk, and by transverse structural frames at the midpoint. Each quadrant had provisions for two inboard and two outboard racks for bombs of 500 pounds or smaller. There were 48 shackling stations, three stations per rack, but only two of the three stations on the outboard racks could be used at one time. Rack positions were changed for 1,000-pound bombs, leaving only one inboard and one outboard rack in each quadrant, and special racks had to be installed to handle 2,000 and 4,000-pound bombs. There was no "snap-open" feature on the B-32's roll-up bomb bay doors, and it took some 13 seconds for the four hydraulic motors to open or close the doors. There was an airspeed reduction of approximately nine miles per hour when the doors were in the full-open position, and limit switches on each door prevented bomb release until the door had reached that position. Maximum rated bomb load was 20,000 pounds.

Oxygen System: The B-32 and TB-32 oxygen system was of the low-pressure demand type, and was supplied by 13 G-1 and 14 F-2 cylinders with provisions for nine additional G-1 units. The G-1 cylinders were installed in the upper portion of the aft bomb bay and under the flooring of the rear crew compartment just forward of the lower ball turret. The F-2 cylinders were installed along the right wall of the rear crew compartment behind the soundproofing panels. Crew members used either the A-10A or A-14 demand-type masks at their duty positions, and there were also five or more portable A-4 or A-6 walk-around bottles. These units were rechargeable at five locations: the bombardier's compartment, two places on the flight deck, and one location each in the rear crew compartment and tail section. Oxygen duration for eight men at 25,000 feet was about ten hours, or eight hours for a ten man crew.

Emergency Egress: Each member of a B-32 or TB-32 crew stowed his parachute near his duty station. The primary escape route for all crew members was through the bomb bay, if this route was possible. The camera hatch in the floor of the aft crew compartment was considered the secondary exit for crew members in that area, while the nose-wheel well was the secondary flight deck exit only if the nose gear could be lowered and locked. If this could not be accomplished, flight deck personnel were to exit the aircraft either through the bombardier's compartment or the camera hatch. Two static lines were provided for the bailing out of wounded or unconscious crew members, one on the right side of the flight deck floor and the other near the camera hatch.

The B-32, like all military aircraft, underwent a series of modifica-

tions throughout its service life. Most of these modifications resulted from suggestions made by the Army Air Force agencies which carried out the flight testing of production B-32 aircraft. These agencies, the Army Air Force Proving Ground Command (AAFPGC), the Army Air Force Air Technical Service Command (ATSC) and the Army Air Forces Tactical Center (AAFTAC) carried out extensive tests of the B-32 in 1944 and 1945. Their recommendations resulted in many changes to the internal equipment of the B-32, as did the suggestions of combat crews that operated the aircraft in the Pacific during the last months of the war. These changes were made to the aircraft by production blocks, and are best examined on a block-to-block basis.

B-32-1-CF (Airplanes 1 through 10, serial numbers 42-108471 to 42-108480): As mentioned above, the first production B-32 was originally fitted with a 16½-foot B-29 tail assembly, but was later equipped with the 19½-foot fin that characterized all subsequent B-32 aircraft. The revisions made to aircraft in this block were: 1) changes to the magneto cooling and addition of a magneto pressurizing system; 2) the change of aileron control wheel travel from 135 degrees to 150 degrees; 3) incorporation of elevator tabs having cambered lower surfaces for increased stability; 4) modification of the hydraulic uplatch system on the main landing gear; 5) revision of the ammunition feed system to nose and tail ball turrets; 6) installation of VHF SCR-522A Radio Set and SCR-274 Range Receiver in lieu of SCR-274 Command Set; 7) installation of the AN/ARC-8 Liaison Radio Set; 8) installation of AN/AIC-2 Interphone System; 9) installation of AN/ARN-7 Radio Compass in lieu of the SCR-296-G system; 10) installation of AN/ARN-5 and RC-103 Instrument Approach System; 11) installation of AN/APN-4 Loran Radio Navigation Equipment; 12) installation of AN/APN-1 Radar Altimeter; 13) installation of SCR-729 Interrogator-Responder (IFF) Set; 14) installation of AN/APQ-5 Low Altitude Bombing Set and AN/APQ-13 Radar Bombing and Search Set; 15) redesign of the radio operator's table; 16) installation of combined radar/navigator operator's table; and, 17) the installation of additional inverters and generators for radar equipment.

B-32-5-CF (Airplanes 11 through 14, serials 42-108481 through -108484): Adoption of twin rudder tabs and modified engine fire seals as described for TB-32-5-CF, rest the same as B-32-1-CF, except that 42-108482 was special. See below.

TB-32-5-CF (Airplanes 15 through 25, serials 42-108485 to 495): Consolidated-Vultee's contract with the Army called for the construction of 40 TB-32 aircraft, and production began in the 5-CF block. Major revisions included: 1) deletion of all offensive and

This view from above, probably of 42-108476, shows the Dominator's 135-foot Davis wing to good advantage. The narrow-chord wing had comparatively low drag, a stable center of pressure, and developed lift at relatively small angles of attack. One source of trouble on the production aircraft was the trailing-edge-over-flap panels, which were found to have developed cracks during the service and combat tests. The recommended field fix was to replace the panels with thick 24ST aluminum sheets.

USAF

defensive armament, fairing over of turret cut-outs and addition of 750 pounds of ballast; 2) deletion of the AN/APN-4 Loran Set, AN/APN-1 Radar Altimeter, SCR-729 Interrogator-Responder (IFF) Set, AN/APQ-5 Low Altitude Bombing Set and AN/APQ-13 Radar Bombing and Search Set; and, 3) revision of the engine mount ring inner diaphragm to accept a new engine fire seal adapter flange on the R-3350-23A engines.

TB-32-10-CF (Airplanes 26 through 50, serials 42-108496 to 520): The second block of TB-32s featured, 1) a redesign of the bombardier's entrance door to facilitate single-operation jettisoning; 2) installation of engine fire extinguishers; and, 3) replacement of SCR-269-G Radio Compass by the AN/ARN-7 set.

TB-32-15-CF (Airplanes 51 through 54, serials 42-108521 to 524): The final block of TB-32s received, 1) a redesigned glide-bombing system; and 2) standardized installation of tail member deicer boots.

B-32-20-CF (Airplanes 55, 56 and 58 through 75, serials 42-108525, 526 and 528 through 545): This block marked the return to production of combat-equipped aircraft following the completion of the TB-32 contract. These aircraft, and all subsequent B-32s, retained the modifications made to the two preceeding blocks of TB-32 airplanes as well as 1) the re-installation of all offensive and defensive armament; 2) re-installation of full combat radio and radar suite; 3) installation of flak blankets; 4) addition of simplified and improved bomb-hoisting equipment; 5) the addition of improved ditching equipment in the form of life raft stowage compartments on top of the fuselage just forward of the aft turret; 6) installation of cargo-carrying platforms in bomb bays; 7) installation of cowl flap position indicators; 8) replacement of B-10 bomb shackles by B-7 units; 9) installation of jettisonable cockpit side windows; 10) replacement of B-3 driftmeter by AN-5763-40 unit; and, 11) addition of scanning blisters on rear fuselage to permit visual inspection of engine nacelles and main landing gear.

B-32-21-CF (Airplane 57, serial 42-108527): The third Fort Worth-built B-32-20 was held at the factory for conversion into an experimental paratroop-carrier. The aircraft was stripped of all bombing equipment and fitted with "bleacher" type seats in the aft crew compartment and aft bomb bay. The paratroopers were apparently supposed to exit the aircraft through the forward bay. This proposal was not followed up, however, and the aircraft was held at the factory until declared excess in October 1945.

B-32-25-CF (Airplanes 76 through 100, serials 42-108546 to

B-32 pilot's compartment. This view shows the main instrument panel, upper panel, and pedestal controls, with the crawlway to the bombardier's compartment and nose turret just visible between the pedestals.
USAF/AFSHRC

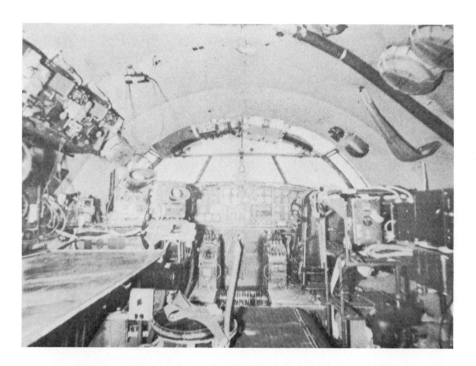

Top: The B-32's flight deck was relatively spacious by contemporary standards. USAF/AFSHRC

Bottom left: Detail of pilot's position on B-32-25-C 42-108547. Jettisonable side windows had swing-open Plexiglas ports. The instrument mounted above the panel behind the words "Radio Compass Indicator" was a compass used as part of the service test special instrumentation. USAF/AFSHRC

Bottom right: Radio operator's position, on radar-equipped aircraft, was on the starboard side of the flight deck just aft of the co-pilot's seat. USAF/AFSHRC

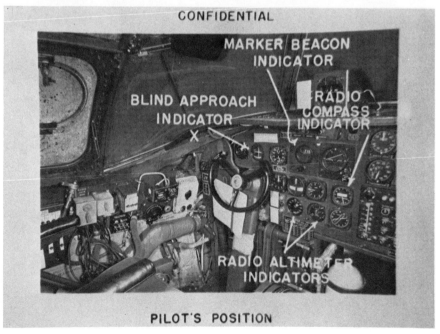

CONFIDENTIAL

MARKER BEACON
INDICATOR

BLIND APPROACH
INDICATOR
X

RADIO
COMPASS
INDICATOR

RADIO ALTIMETER
INDICATORS

PILOT'S POSITION

T-47/ART-13
TRANSMITTER

-17-

-108570): The revisions to this block of aircraft included: 1) the provision of two 750 U.S. gallon self-sealing droppable bomb bay fuel tanks, which increased the total fuel capacity to 6,960 U.S. gallons; 2) installation of the AN/APN-9 Loran Set in lieu of the AN/APN-4 unit; and, 3) the addition of gas-tight seals to the lower surface of the wing sections nearest the fuselage.

B-32-30-CF (Airplanes 101 through 107, serials 42-108571 to -108577): These aircraft received the following modifications: 1) the replacement of AN/APQ-13 Radar Bombing and Search equipment by the AN/APQ-13A set; 2) installation of AN/APQ-2, AN/APT-1 and AN/APT-2 Radar Countermeasures equipment; 3) installation of Sperry A-17A stabilized nose and tail turrets in place of the former A-17 unstabilized models; and, 4) the reinforcement of the wing trailing edge to eliminate the tendency to crack found in some earlier aircraft.

B-32-35-CF (Airplanes 108 through 114, serials 42-108578 to -108584): As for B-32-30-CF, except for addition of extra ammunition for guns.

BC-32-20-CO (Airplane 115, serial 44-90486): This was the one production airplane completed by Consolidated Vultee's plant in San Diego. The 1943 contracts called for 500 planes from the California factory, but only this plane was delivered (28 July 1945) before all production was halted.

B-32-5-CF (Airplane 12, serial 42-108482): Consolidated modified this plane to accept the Sperry A-17A stabilized turrets for testing. The AAF also had it fitted with the four-gun A-18 dorsal turrets and the four-gun A-19 ventral ball. Both were large—but very cramped—manned units mounting two .50 cal. guns in a stacked pattern on each side of the gunner, instead of the usual one gun per side.

B-32-20-CF (Airplane 65, serial 42-108535): Test plane for the "Albert" offensive armament. Its bomb bay was modified to carry the 12,000-lb. *Tall Boy* bombs, requiring relocation of certain antennas.

B-32 Special Photographic Aircraft (serials unknown): The AAF outfitted three B-32s with large numbers of cameras for photomapping Japan. Combat crews were being assembled at Wright and Patterson Fields when the dropping of the atom bombs cancelled this mission and ended the war.

B-32-40-CF (Airplane 171, serial 42-108641): This and subsequent planes were to have modifications to improve the bombardier's forward vision, details unknown. Aircraft never completed.

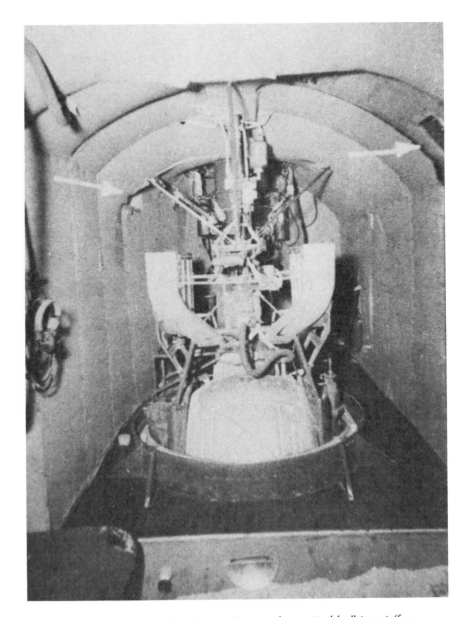

Aft compartment, looking forward toward retracted ball turret (foreground) and rearmost dorsal turret. Arrows point out a small window (left) and an escape hatch (right). USAF/AFSHRC

Navigator's position on radar-equipped aircraft, just aft of the pilot's seat, showing LORAN and SCR-729 installations. USAF/AFSHRC

Flight deck, looking aft from cockpit. The mount for the forward dorsal turret is at center-rear. USAF/AFSHRC

Dominator tail development.

A. *XB-32 Numbers 1 and 2.*

B. *XB-32 Number 3 and first B-32-1-CF (42-108471).*

C. *B-32-1-CF.*

D. *B-32-5-CF, TB-32-5, -10 and -15-CF.*

E. *TB-32-15-CF, B-32-20-CF and all subsequent.*

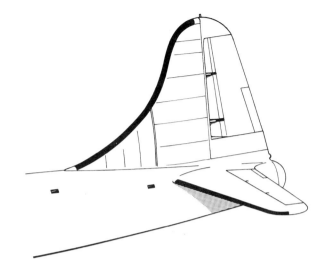

B.

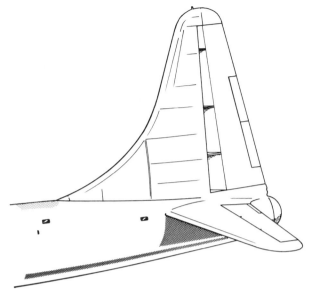

D.

C.

E.

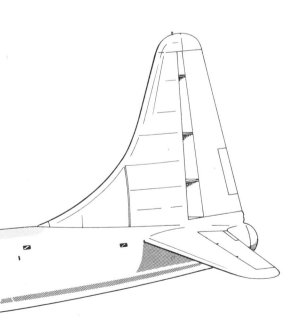

A.

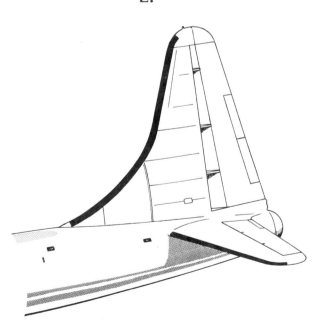

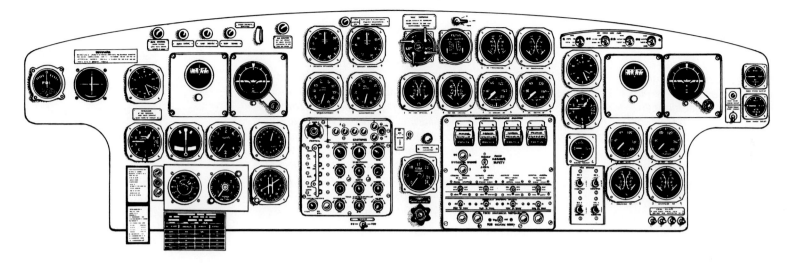

Main instrument panel.

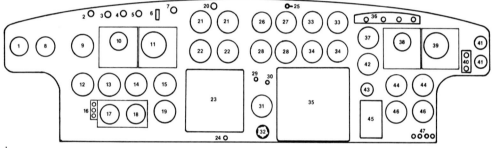

Control and indicator identification key.

1. Pilot's Direction Indicator (PDI)
2. Ball turret warning light
3. Radio marker beacon
4. Bomb doors warning light
5. Bomb release warning light
6. Bomb salvo switch
7. Salvo switch warning light
8. Blind approach indicator
9. Airspeed indicator
10. Directional gyro
11. Flight indicator
12. Altimeter
13. Turn-and-bank indicator
14. Rate-of-climb indicator
15. Radio compass
16. Altitude limit indicator, AN/APN-1
17. Altitude limit switch, AN/APN-1

18. Altitude indicator, AN/APN-1
19. Flux gate compass repeater
20. Both-inverters-out warning light
21. Manifold pressure gauges
22. Tachometers
23. C-1 automatic pilot control panel
24. Alarm bell switch
25. Flaps switch
26. Ignition switches
27. Flap position indicator
28. Cylinder head temperature gauges
29. Landing gear switch
30. Landing gear down-lock light
31. Master tachometer, propeller system
32. Proportional syncho-control knob,
    propeller system

33. Oil temperature gauges
34. Main oil pressure gauges
35. Propeller system control panel
36. Oil cooler exit flap switches
37. Airspeed indicator
38. Directional gyro
39. Flight indicator
40. Hydraulic pump override switch
    and light
41. Brake hydraulic pressure gauges
42. Altimeter
43. Main hydraulic pressure gauge
44. Nose oil pressure gauges
45. Intercooler switches
46. Carburetor temperature gauges
47. Cowl flap switches

The main gear of the B-32 consisted of four 56-inch tires and tubes and four wheel and brake assemblies, all of which were interchangeable with those of the B-24. The gear retracted hydraulically up and forward into wells in the inboard nacelles in about ten seconds. Rods with universal joints pulled the fairing doors closed to completely cover the retracted gear.

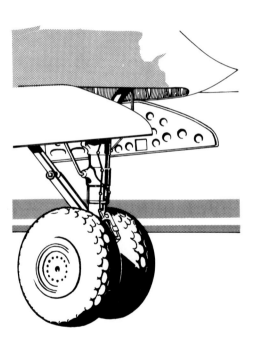

The B-32's outer wing panels were similar to those of the B-24, there being a slight difference in the contour of the wing tip and outboard end of the aileron. The aileron trim tabs were operated and controlled electrically by a motor housed in each aileron. The tab balance (servo) deflection was mechanically produced by the control linkage, in order to take some of the load off the controls.

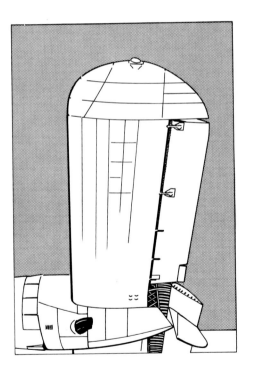

The propeller system on the B-32 was completely electric, controlled by switches on the instrument panel. The flight crew could select the desired engine speed by manually setting a master tachometer, then the propeller control automatically maintained that speed by adjusting blade pitch, thus achieving automatic synchronization of all engines and all propellers. Under manual switch control, the system also provided fixed pitch, reverse pitch on inboard props, and feathering. The hub assembly consisted of a "spider" machined from a solid steel alloy forging and served as a mount for six slip rings, an alcohol slinger ring, the four blade assemblies, and the power unit. Each hollow steel blade was fitted with an aluminum alloy cuff to improve engine cooling, particularly during ground operation. The power unit consisted of a reversible electric motor, speed reducer, power gear assembly, and cam-operated limiter switches.

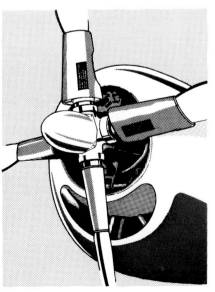

Main gear extension stages.

Nose gear extension stages.

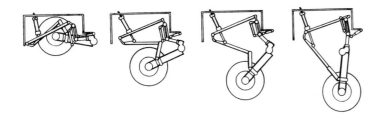

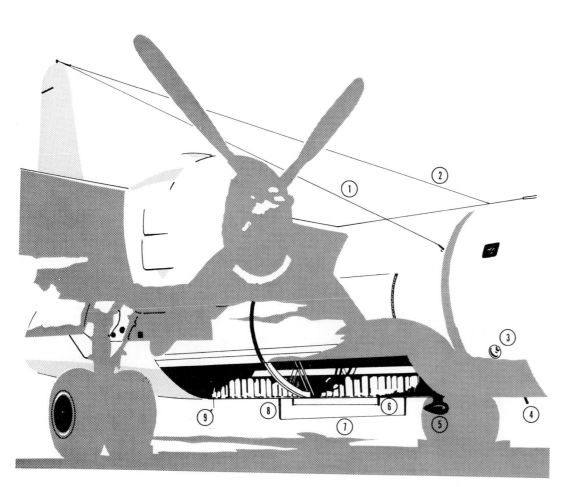

The Dominator's two bomb bays were divided into left and right sections by a large built-up beam that provided structural strength and carried a catwalk between fore and aft compartments. Each of the four sections was large enough to accommodate a single 4000-lb. bomb or a removable self-sealing fuel tank of approximately 750 U.S. gallons capacity. On planes equipped with radar, the two wire antennas above the fuselage were for the liaison radio set (1) and the range receiver (2). Antennas for the radio compass receiver (5), the marker beacon receiver (6), radio compass sense (7), the VHF radio set (8), and the IFF set (9) were on the underside of the fuselage along the main beam. The mast and weight (4) of a 250-foot reel antenna protruded from the lower left section of the forward compartment. Retractable landing lights (3), one on each side, were controlled by two 3-position switches mounted on the co-pilot's control pedestal.

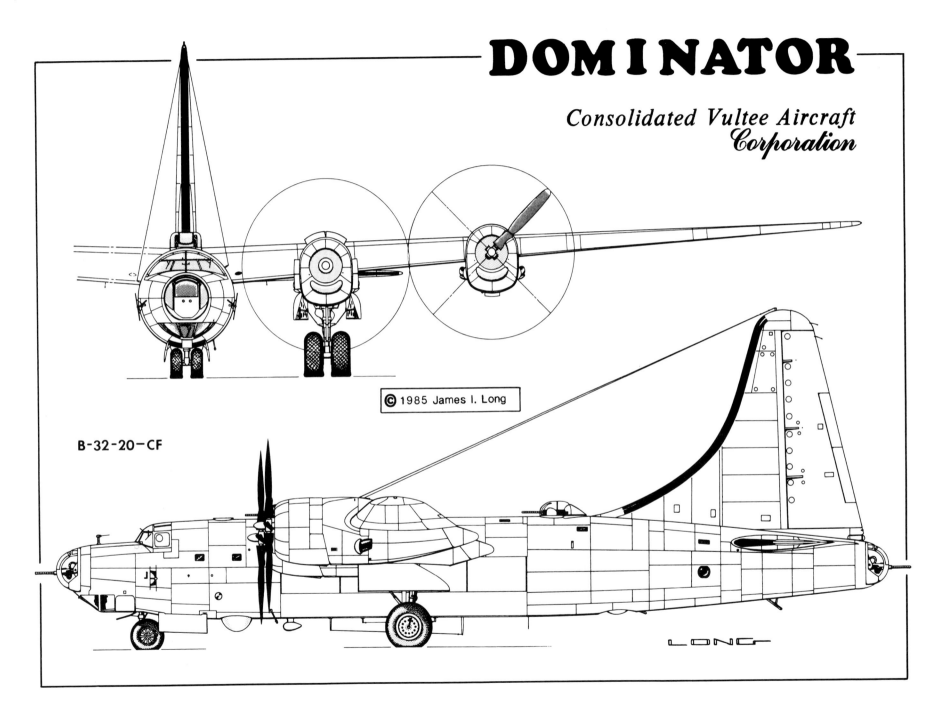

DOMINATOR

Consolidated Vultee Aircraft Corporation

© 1985 James I. Long

B-32-20-CF

LONG

CHAPTER FOUR
Testing the B-32

The Army Air Forces had originally intended to deploy the B-32 for combat operations by the summer of 1944 as part of the "48 Very Heavy Group Program." According to this plan, several B-24 groups in the Mediterranean Theater of Operations (MTO) would convert to B-32s as soon as the aircraft's service tests ended. Upon satisfactory completion of this conversion the remaining B-17 and B-24 groups in the MTO and European Theater (ETO) would trade their aircraft for B-32s, thus giving the Eighth and Fifteenth Air Forces the long range striking power their Flying Fortresses and Liberators did not offer. But by the spring of 1944 it was clear that the continuing delays in the B-32 program would make it impossible to implement the "48 Group" plan on schedule, for the essential pre-deployment flight testing of the B-32 had not yet even begun. Indeed, the official AAF directive concerning the functional and tactical testing of production B-32 aircraft was not issued by the Materiel Command at Wright Field until 15 August 1944, during which month the AAF officially changed the B-32's name from "Terminator" to "Dominator."

The test directive reflected the Army's desire to put the B-32 through a painfully thorough test program even if it further delayed the actual deployment of much-needed combat aircraft. The Directive called for a program of at least 200 flying hours, with evaluations of the B-32's flight characteristics, operational suitability, armament installations, bombing capabilities, maintenance and servicing requirements and internal equipment. The tests were to be conducted by the Army Air Forces Proving Ground Command (AAFPGC) at Eglin Field, Florida, the Army Air Forces Tactical Center (AAFTAC) at Pinecastle Field, Florida, and the Air Technical Service Command (ATSC) installations at Wright Field, San Diego, Fort Worth, and Vandalia, Ohio. In addition, the directive required field tests of the B-32's combat performance and its suitability for long-range weather reconnaissance.

The Army intended to use all ten production B-32-1-CF aircraft in the service test program, but the problems that had plagued the Dominator since its inception once again disrupted the best-laid plans. The first B-32 delivered to and accepted by the AAF, 42-108472, was wrecked beyond salvage when its nose gear collapsed on landing at Fort Worth. Production delays at the plant held up delivery of the next aircraft, 42-108475, until 22 November. These same problems affected the rest of the first ten production aircraft, and by 31 December 1944 only five B-32s had been accepted at the various test centers.

The slowness of B-32 production and delivery had, by mid-December 1944, once again put the Dominator program in danger of cancellation. The airplane's critics pointed out that the first B-29 combat mission had been flown the previous June, while the AAF had yet to begin serious flight testing of the B-32. Further, those B-32s accepted to date were experiencing an unusually high rate of mechanical malfunctions, and some AAF agencies were beginning to complain of faulty workmanship on delivered aircraft. The Dominator's critics concluded that the implementation of planned B-32 flight crew conversion training would hinder the deployment of combat-ready crewmen for the already proven B-29, and recommended that the Dominator program be cancelled and crews scheduled for B-32 training instead be allocated to operational Superfortress units.

In an attempt to clarify the B-32 situation, Headquarters, USAAF, directed Brigadier General Donald Wilson to undertake an exhaustive analysis of the aircraft's continuing problems. General Wilson's report, submitted on 14 December 1944, quite literally saved the Dominator from immediate extinction. Though the report acknowledged the many difficulties that had plagued the B-32 program, it objectively pointed out that the Army could not financially or militarily afford to abandon the Dominator until a completed service test program had positively shown it to be an unacceptable bombardment aircraft. And, since there was always

The second production article, 42-108472, was scheduled to participate in the B-32 service tests, but crashed at Fort Worth. The aircraft was, needless to say, a total write-off.
USAF via William T. Y'Blood

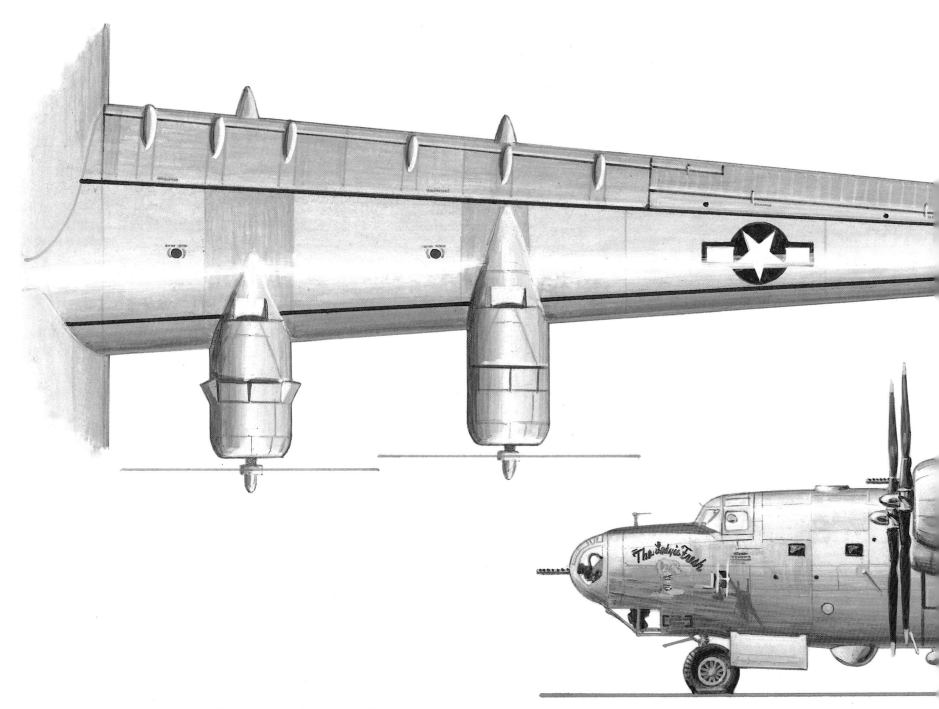

DOMINATOR
Consolidated
B-32 BOMBER

B-32-5-CF 42-108484, field number 580, at Eglin during the service tests.　　　USAF/AFSHRC

Head-on view of 484 taken at Eglin. Note the size of the wheel assemblies and the huge Curtiss props.

USAF/AFSHRC

the chance the B-32 would rise above its teething troubles, the AAF could not afford to cancel the planned crew-conversion training program. General Wilson thus recommended that no final decision be made concerning the cancellation of the B-32 until the service test program was completed, and that crew training be started immediately rather than being postponed until a final determination had been made about the Dominator's combat suitability.

Headquarters, USAAF, accepted General Wilson's recommendations and on 31 December requested the primary testing agency, AAFPGC, to submit a preliminary report on the aircraft's operational limits by 15 February 1945. This deadline soon had to be extended, however, since all B-32-1-CF aircraft thus far delivered to the AAF had gone to ATSC, and AAFPGC had no Dominators to test. Indeed, AAFPGC did not receive its first B-32, 42-108477, until 30 January 1945. A second aircraft, B-32-5-CF 42-108484, reached the command on 22 February, and these two articles formed the backbone of the AAFPGC test program until joined by a B-32-25-CF, 42-108547, in May. The ATSC took delivery of a further 17 Dominators prior to October 1945, including one additional B-32-1-CF (42-108478), two B-32-5-CF (42-108481 and 482), seven B-32-20-CF (42-108525, 526, 533, 535 and 540 through 542), one B-32-25-CF (42-108546), and six B-32-30-CF (42-108571 through 576). The AAF Tactical Center was allocated one B-32-1-CF (42-108480), and four B-32-20-CF (42-108534 through 538).

The B-32 service test program revealed numerous problems with the Dominator's internal equipment and arrangement, including extremely high flight deck noise levels, inconveniently sited instruments and controls, inadequate mechanical subsystems and, most seriously, severe engine nacelle deficiencies which often resulted in powerplant fires. In addition, several AAF agencies complained of poor workmanship and faulty construction methods in aircraft that had already been inspected and officially accepted. The persistent nature of these problems, and the need to correct them before flight tests could be resumed, resulted in abnormally high aircraft out-of-commission rates. The first two Dominators allocated to AAFPGC well illustrate this problem. The aircraft, exclusive of the time they were inactive due to the installation of necessary test equipment, had respective out-of-commission rates of 82 and 59 percent.

The majority of the problems encountered during the B-32 service tests were later eliminated in subsequent blocks of production aircraft, either through design modification or improved quality control, and were balanced to a certain extent by the

RELOCATED ANTENNAS ON AIRPLANE NO. 42-108535

Detail of experimental antenna installations on 42-108535 during the service tests of the "Albert" offensive armament. USAF/AFSHRC

-28-

Dominator's positive characteristics. The aircraft was found to have excellent low speed directional control, good take off and landing manners, and rapid control response. In addition, the B-32 was found to be a stable bombing platform, its definitive defensive armament array offered good all-around protection, its subsystems were easily accessible for maintenance, and its unique reversible inboard propellers gave it excellent ground-handling characteristics unmatched by any comparable aircraft.

Service testing of the B-32 continued throughout the airplane's short military career, though several important tests had to be cancelled due to a lack of available production aircraft. Those tests actually completed were the "Test of Multiple Suspension of Bombs" (July 1945), "Operational Suitability of B-32 Bombing Equipment" (July 1945), "B-32 AN/APQ-13 Installation Suitability Test" (July 1945), "Operational Suitability of the B-32 Aircraft" (August 1945), "Operational Suitability of B-32 Communication Equipment" (September 1945), and "Photographic Suitability of the B-32 Aircraft" (June 1945). All further testing was suspended by Headquarters, USAAF, upon cancellation of the B-32 program in October 1945.

The "Test of Multiple Suspension of Bombs in the B-32 Airplane" was one of the few Dominator tests actually completed before the test program was cancelled. This photo shows 100-pound GP bombs triple suspended on T-15 cluster adapters in one of the service test aircraft. USAF/AFSHRC

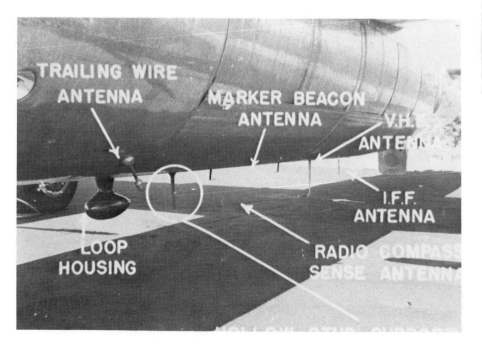

Revised antenna installations on 42-108484. This arrangement became standard. USAF/AFSHRC

Unloading 100-pound GP bombs during the Eglin Field service tests. USAF/AFSHRC

42-108484 drops sextuple-suspended M47A2 incendiary clusters in train at 1/10 second intervals during the bombing suitability segment of the service tests. USAF/AFSHRC

CHAPTER FIVE
Training

In the fall of 1944 Headquarters, USAAF, had directed the AAF Training Command to institute training courses for prospective B-32 flight and ground crews despite the very real possibility that the the entire Dominator program might be cancelled at any time. Training Command moved ahead with the necessary planning despite the uncertainties, and the 3704th Army Air Forces Base Unit (AAFBU) at Mississippi's Keesler Army Air Field activated the B-32 Airplane and Engine Mechanic School in October 1944. Other technical schools soon followed, including the 3502nd AAFBU at Chanute Field, Illinois (engine training), and the 3712th AAFBU at San Diego's Lindbergh Field (a Consolidated-run maintenance training school).

For the first few months of their existence these B-32 technical schools were plagued by a shortage of necessary training aids. The Keesler facility, for example, started operations with only three obsolete R-3350 engines, one propeller synchronizer, a few early B-32 technical manuals and a single factory catalog. Things improved somewhat for the technical schools following General Donald Wilson's December 1944 analysis of the B-32 problem, though it wasn't until the spring of 1945 that the Dominator technical training program really hit its stride.

The B-32 Flight Crew Transition Training School, activated by the 2519th AAFBU at Fort Worth Army Air Field in October 1944, originally suffered from the same shortage of equipment that hindered the technical schools. The 2519th did not receive its first Dominator, a TB-32-5-CF (42-108489), until 27 January 1945. This situation began to change, however, once greater numbers of production aircraft became available. Between 27 January and 6 July 1945 the 2519th received a further nine TB-32-5-CF aircraft (serials 42-108485, 487, 488 and 490 through 495), 25 TB-32-10-CF (42-108496 through 520), four TB-32-15-CF (42-108521 through 524), 18 B-32-25-CF (42-108551, 553 and 555 through 570) and one B-32-20-CO (44-90486).

The B-32 Flight Crew Transition School at Fort Worth was designed to provide combat-ready Dominator crews in as short a time as possible. The vast majority of the pilots, co-pilots, bombardiers, navigators, radar observers, aerial engineers and gunners assigned to the transition course had already qualified on the B-24, and thus needed only a two-month transition course to convert to the B-32. The two-month period was broken into two equal parts, during the first of which the crew members were dispersed to specialized schools. The commander/pilot, co-pilot and aerial engineer studied together at Fort Worth; the bombardier, navigator and radar observer went as a group to other fields, and the gunners learned about the Dominator's defenses at Laredo Army Air Field, Texas. During the first phase of training prospective B-32 aircraft commander/pilots underwent 50 hours of transition time in a TB-32, while co-pilots received 25 hours of flight time and 25 hours of observer time. In the second month of training the entire crew reassembled at Fort Worth for 80 hours of group flight time, 40 hours in TB-32s and the remainder in combat-equipped B-32s.

According to AAF Training Command's original plan, crews that had completed their transition training at Fort Worth would be assigned to the Fourth Air Force for final phase training and staging. The 426th AAFBU at Mountain Home Army Air Field, Idaho, was the unit chosen for this duty. The 426th received its first Dominators, B-32-25-CF 42-108548 and 549, on 26 May 1945. Two more aircraft from the same production block, 42-108550 and 552, arrived four days later, and were followed on 6 June by 42-108554. However, the Army once again changed its plans, and all five B-32s at Mountain Home were re-assigned to the 2519th AAFBU at Fort Worth on 27 June.

Like all other aspects of the Dominator's military career, the B-32 flight crew transition training program was hindered by a myriad of problems. Besides the almost routine shortage of aircraft and training aids, the program at Fort Worth had to contend with the Dominator's always tempermental mechanical systems. During April 1945, for example, three TB-32s were severely damaged when their main gear collapsed on landing. One of the accidents was due to crew error—the flight engineer inadvertantly hit the gear switch while trying to raise the flaps on 42-108488—but the other two (42-108487 and 510) were caused by faulty main gear shock-strut locking assemblies. Main gear failures damaged two more aircraft (42-108499 and 531) during the first week in May, whereupon all B-and TB-32 aircraft were grounded. Within 20 days all Dominators had been fitted with

The first Dominator received at Fort Worth Army Airfield, TB-32 42-108489, taxis across the runway from the Consolidated-Vultee plant on 27 January 1945.

Robert A. Garfield

Faired-over dorsal turret cutouts are plainly visible in this above shot of a TB-32. The black lines near the leading and trailing edges of the main wing are two-inch wide walkway borders. The round window aft of the cockpit is the navigator's astro glass, which consisted of a fixed-in-place, optically flat plate rather than a retractable Plexiglass dome.   USAF modified down lock latches designed by the Bendix Corporation, but valuable training time had been lost.

A more serious problem facing the Fort Worth program was that of Dominator engine fires. On 8 March 1945 a TB-32-5-CF, 42-108495, experienced a fire in number three engine during a routine training mission. The fire could not be extinguished, and the crew was forced to abandon the flaming aircraft. All personnel survived the incident, but their Dominator was a total write-off. Two days later a B-32-1-CF (42-108475) assigned to the ATSC facility at Fort Worth also crashed due to an uncontrollable engine fire. A rash of less serious fires damaged several other aircraft, and a subsequent investigation showed the cause of the problem to be faulty engine fire seal adapter flanges. The deficiency was corrected by a revision to the engine mount ring on all aircraft, and led to the installation of engine fire extinguishers on all later produc-tion models.

The B-32 underwent an unending series of design changes and modifications during its service career, often as many as 400 in a single week, and this too hindered the Fort Worth training program. On numerous occasions the 2519th AAFBU Supply and Maintenance Branch ordered replacement parts for its aircraft, only to be told the part was no longer being made in the same size and shape or had been discontinued entirely. As an example, the TB-32-5-CF damaged by a nose gear collapse on 25 April, 42-108488, required a complete replacement bombardier's compartment bulkhead. The part was no longer manufactured to the same specifications, and the Dominator remained inactive until Consolidated could set up a new jig to build the item in the required size. This was not an isolated incident either, for the official history of the 2519th AAFBU complained that "it recurres [sic]

This TB-32 rear view shows the faired-over tail ball position to good advantage. Note the open rear belly hatch and the ex-tended tail skid.
General Dynamics

practically every day on small parts which are no longer available in their original 'hand-made shapes' for early B-32's.'' And, though the base engineering shops were able to fabricate some of the needed parts, many were simply too large for local manufacture. To further complicate this supply and maintenance nightmare, many of the parts that were available bore entirely different numbers from those in the supply catalogs.

Despite these problems the various B-32 Dominator training centers managed to supply a steady stream of qualified aircrew and maintenance personnel. By the time Headquarters, USAAF, cancelled the Dominator program in October 1945, Training Command had turned out some 240 pilots and co-pilots, over 140 aerial engineers, and comparable numbers of radar observers, navigators, gunners and mechanics.

One mission of the B-32 school at Fort Worth was to train fourteen demonstration crews from the Fourth Air Force, one of which is seen here. In the front row, left to right, are Sgt. Billy Robertson (engineer), Sgt. Charles Foreman (engineer-gunner), Sgt. Lloyd O'Morrow (radio operator), Sgt. Paul Hatt (armorer-gunner), Cpl. Orville Fruetche (gunner), and Cpl. Walter Radomski (gunner). Standing behind, left to right, are 1st Lt. Christopher Cammack (pilot), 2nd Lt. Earl R. Meisenheimer (co-pilot), 2nd Lt. Earl H. Smith (bombardier), and 2nd Lt. Emanuel Lipton (navigator). The aircraft providing the background for this photo is 42-108574, a B-32-30-CF. Upon delivery in June 1945, 574 was assigned to the specialized depot at Topeka, Kansas, probably for the installation of RCM equipment (an RCM blade antenna can be seen on the lower fuselage directly below the forward turret). After that the aircraft was sent to the Air Proving Ground at Eglin Field, where it picked up the field number (596) that it still bore when this photo was taken at Fort Worth, the plane having been reassigned there in August 1945.
Earl R. Meisenheimer

Forty TB-32s were built, primarily to equip the pilot transition school at Fort Worth. The plane in the foreground is a TB-32-5-CF, 42-108490. Note the open bombardier's hatch just forward of the nose wheel.

General Dynamics via William T. Larkins

The ground crew stands clear as 42-108522, a TB-32-15-CF, starts her engines pior to a training mission. The field number OM12 was originally assigned to TB-32-5-CF 42-108488, but that aircraft went out of service in July 1945 and the number was given to 522.

USAF via William T. Y'Blood

Sperry A-17 nose and tail turrets in the armament training area at Fort Worth. The unit second from right shows the rear entrance hatch. USAF/AFSHRC

The Dominator was not officially shown to the public until after its combat debut in the Pacific. The first public display of the aircraft, shown here, took place on Air Force Day, August 1945. The aircraft is TB-32-10-CF 42-108514. USAF/AFSHRC

This in-flight shot of 42-108522 shows to good advantage the sleek lines of the TB-32s. The factory installed some 700 pounds of ballast to compensate for the absence of turrets.

General Dynamics

CHAPTER SIX
Combat

The B-32 test directive issued by the AAF in August 1944 had stipulated that a test of the Dominator's combat suitability be made prior to the B-32's introduction into service units. However, the continued controversy surrounding the B-32 program had, by April 1945, made the accomplishment of such a test unlikely. The AAF Proving Ground Command was the only agency authorized to conduct such a test, and that agency's experience with the B-32 led it to oppose both a combat test and the general introduction of the aircraft into combat units. It thus seemed, in the spring of 1945, that the Dominator was to spend the rest of the war in the same operational limbo it had occupied throughout its troubled life.

But this situation changed abruptly in March 1945 with the arrival in Washington of Lieutenant General George C. Kenney. Kenney, the Commander of the Far East Air Forces (FEAF), had long believed that very long range bombers flying from airfields within his command could deal Japan the knockout blow needed to end the war in the Pacific. He had thus followed the development of the B-29 with extreme interest, hoping that the Superfortress could be used by FEAF to attack the vital Japanese targets in the Dutch East Indies that had so far been beyond the range of American aircraft. His repeated requests for B-29s had been continually denied, however, on the grounds that the huge bombers were badly needed elsewhere. A less determined individual might have given up in the face of such continued opposition, but Kenney

simply switched tactics. As soon as it became obvious that no Superfortresses would be forthcoming, Kenney stopped asking for them and started asking for B-32s.

The purpose of General Kenney's March 1945 trip to Washington, though officially for general talks on the conduct of the air war in the Far East, was thus actually to secure B-32s for FEAF. To further this aim Kenney and his staff met with General H.H. Arnold, commanding general of the USAAF, beginning on 15 March. Two combat-equipped Dominators were flown to Washington on 18 March, 42-108477 from Eglin Field and 42-108478 from Wright Field, to assist the senior officers in their deliberations. After several days of conferences, demonstration flights and staff briefings, Kenney officially requested the assignment of "several" Dominators to FEAF "for testing purposes."

In his request for the B-32s General Kenney pointed out that, despite the problems thus far encountered with the aircraft, Dominators were still being produced and combat crews were still being trained. It would be a waste of valuable, *immediately available* resources not to use the aircraft and crews in some way, even if only in a limited role. Further, Kenney argued that the deployment of combat-equipped B-32s to FEAF would serve as the acid test of the Dominator's operational suitability. If the aircraft proved itself in combat the money spent in producing it and training crews would not have been wasted, and plans could then be made to deploy the B-32 in large numbers. If, on the other hand, the Dominator was not equal to the task, it could be at last cancelled for good and the men and money invested in the program could be shifted to the B-29. Finally, Kenney argued that no matter how well the B-32 performed it would still put pressure on the Japanese and thus make at least a small contribution to the war effort.

General Arnold had been fairly impressed by the B-32 and considered it a definite improvement over the B-17 and B-24 despite its teething troubles. After considering General Kenney's recommendations, Arnold decided on 27 March to authorize a comprehensive B-32 combat test. He explained his decision by saying:

We are trying to win this war in the shortest possible time. To aid this, the B-32 must replace the B-24 in the Pacific in order to bring the maximim striking force agsinst Japan. . . . If that course is followed, the maximum possible bomb load will be brought against the Japs in the shortest possible time....

Events moved swiftly once the decision had been made to deploy the B-32 to the Pacific. General Arnold appointed Colonel Frank R. Cook of ATSC as test detachment commander, and the test itself was designated Special Project 98269-S. A total of 32 individuals were assigned to the test contingent, 12 from AAFPGC, four each from ATSC, AAFTAC and Consolidated-Vultee, three from Air Transport Command, two each from the Army Air Forces Board and AAF Training Command, and one from the Office of the Assistant Chief of Air Staff for Operations, Commitments and Requirements, USAAF. All personnel were to assemble at Consolidated's Fort Worth plant by 1 May for processing and familiarization with the three aircraft assigned to the test. The entire group would then stage with the Dominators to Mather Army Air Field in Southern California, from where the trans-Pacific ferry flight would begin on 16 May.

Ground personnel inspect "Hobo Queen II," 42-108532, soon after her arrival at Clark Field on 24 May 1945. Note the crewman standing in the open navigator's astro glass hatch.
San Diego Aerospace Museum

The three B-32-20-CF aircraft chosen for the combat test were 42-108529, 531 and 532, and were to be flown by, respectively, Colonel Cook, Major Henry S. Britt and Colonel Frank Paul. But a main gear failure caused 531 to crash on completion of a required 12-hour fuel consumption test flight, and Major Britt and his crew were assigned 42-108528 as a replacement. This Dominator had formerly been a test aircraft at Fort Worth and was somewhat the worse for wear. Indeed, Major Britt noted no less than 32 mechanical deficiencies when he took command, and later called 528 the "poorest combat airplane" he had ever flown.

Major Britt's low opinion of his aircraft was to be continually reinforced during the trans-Pacific ferry flight. The other two Dominators, "Hobo Queen II" (532) and "The Lady is Fresh" (529), left Fort Worth as planned on 12 May, but gremlins kept 528 in Texas for two additional days. Then, with all three aircraft set to depart Mather Army Air Field on 16 May, Britt's baulky Dominator threw another mechanical tantrum and had to abort. The other aircraft left on schedule, and from that point on a series of minor malfunctions forced an increasingly frustrated Britt and his crew to play catch-up all the way across the Pacific. The three aircraft made their way to Luzon by way of Hawaii, Kwajalein and Guam, and 528 was at least a day behind the others at each point. "The Lady is Fresh" and "Hobo Queen II" made a relatively uneventful crossing and arrived in the Philippines on 24 May. Major Britt nursed his reluctant bomber into Luzon on the 25th, and all three Dominators remained at Clark Field for the next four days preparing for their combat debut.

The B-32 combat test was conducted under the auspices of General Kenney's Fifth Bomber Command, and the 386th Bombardment Squadron of the 312th Bombardment Group (Light) was chosen as the host unit. The Group had begun its combat life in December 1943 as a New Guinea-based P-40 outfit, converted to A-20 light bombers in February 1945, and moved to Floridablanca, Luzon, the following April. Though the Group Commander, Lieutenant Colonel Selmon Wells, was in nominal charge of the test, he was directed to work "very closely" with the test detachment commander, Colonel Cook. Lieutenant Colonel Wells and his intelligence staff would select suitable targets and provide all necessary support for the combat crews, while Colonel Cook would have final say on which targets would be attacked and which aircraft would be used. Both men were to submit separate weekly reports on the progress of the test, and each would file a summary report upon the conclusion of the test program.

The emblems of the 312th Bombardment Group (Light) and its four squadrons.
Roy G. Cherry

The combat test, as originally planned, was to consist of eleven missions flown in varying weather conditions against as wide a variety of targets as possible, and was to be completed by 1 July. If the B-32 proved itself the 312th Group's remaining squadrons, the 387th, 388th and 389th, would give up their remaining A-20s and convert entirely to B-32s. AAF Headquarters anticipated that, should such a conversion occur, the 312th would be the first of at least three planned Dominator Groups in the FEAF.

The first B-32 combat mission was flown from Clark Field on 29 May against a Japanese supply depot at Antatet in Luzon's Cagayan Valley. All three Dominators were scheduled to take part in the attack, but a broken supercharger forced 528 to abort on takeoff. "Hobo Queen II" and "The Lady is Fresh" departed Clark at 1030 hours, each with a full fuel load and nine 1,000-pound bombs, tested their guns over Baler Bay and set course for Antatet. The two Dominators arrived over the target just before noon and made individual bombing runs from 10,000 feet. The weather was near perfect and there was no interference from flak or fighters, and the Japanese outpost was quickly reduced to smoking rubble. Both aircraft stayed in the area only long enough to take strike photos, then returned to Clark Field without incident.

Top: Colonel Frank R. Cook (third from right) and his crew pose in front of "Hobo Queen II" soon after arrival at Floridablanca, Luzon. Note early version nose art. Lindy LeVine via William T. Y'Blood

Bottom left: Captain William P. Barnes, seen here standing in front of one of the three B-32s that participated in the first raid, conducted the pre-strike crew briefing. When this photo was taken Captain Barnes had just 91 days left to live. He and all others aboard 42-108544 were killed in a takeoff accident on 28 August. The black spots on the center of the nose turret and the engine cowl at left are flaws that were on the original 2 × 3-inch snapshot. USAF/AFSHRC

Bottom right: Captain Barnes conducts a last-minute briefing for Mission 149-A-11 in the scant shade of a Davis wing. The tall man wearing the cap at extreme left is Lieutenant Colonel Selmon Wells, the 312th Group commander at the time. He went along on the raid as an observer. USAF/AFSHRC

Engine starting and warm-up prior to the first raid. 42-108529 (foreground) starts number three engine as "Hobo Queen II" taxis out. Note that only the name "The Lady is Fresh" appeared on the right side of 529's nose.   USAF/AFSHRC

"The Lady is Fresh" makes her way to the end of the runway. She was destined to be the first B-32 to drop bombs on an enemy position. Note C-47s in the background.   USAF/AFSHRC

Two days after the attack on Antatet all three B-32s moved to the 386th's base on Floridablanca. Crews from neighboring B-24, B-25 and B-26 units started arriving almost immediately for transition training, and Colonel Cook's men soon set up a comprehensive technical school to train mechanics, gunners and radio operators. The three B-32s were kept busy with a series of practice missions, though 42-108528 continued to experience mechanical problems and was often grounded for repairs.

These problems kept 528 from participating in the first B-32 strike launched from Floridablanca, which was flown on 12 June against Batan's Basco airfield. "The Lady is Fresh" and "Hobo Queen II," each carrying forty 500-pound bombs, set out at 0930 and arrived over the target 90 minutes later. No opposition was encountered, and each aircraft made two runs. The runway was definitely out of commission when the Dominators left, and both aircraft returned to Floridablanca unscathed.

The remaining nine test missions were flown from Florida-

blanca between 13 and 25 June. The third strike (13 June) marked 528's combat debut, when she and "Hobo Queen II" struck the Japanese airfield at Koshun, Formosa. Mission number four (15 June) put "The Lady is Fresh" and "Hobo Queen II" over the huge sugar refinery at Taito, Formosa, which was obliterated by sixteen 2,000-pound bombs. All three Dominators returned to Taito City the next day on raid number five, giving the area the "Tokyo Treatment" with a total of one hundred and twenty 500-pound incendiary clusters. "Hobo Queen II" carried out the sixth test mission alone on 18 June, when she combined a shipping search of the South China Sea with a night raid on Hainan Island in the Tonkin Gulf. The three B-32s returned to Formosa on 19 June to bomb a series of railway bridges, and "The Lady is Fresh" and 42-108528 paired up for missions eight and nine (20 and 22 June) against other targets on Formosa. The tenth B-32 strike was a single plane shipping search flown by "Hobo Queen II" on 24 June in the Canton-Macao area off the Chinese mainland. "The Lady is Fresh"

Takeoff. The Dominators took to the air at 1030 hours, local time. Ninety minutes later the first bombs fell 10,000 feet onto the Japanese emplacements in the village of Antatet, Cagayan Valley, Luzon. USAF/AFSHRC

joined "Hobo Queen II" for the last test mission, flown on 25 June against railway bridges near Kiirun, Formosa.

During the eleven missions that comprised the B-32 combat test the three Dominators flew a total of 20 sorties and dropped nearly 134 tons of bombs. The only serious opposition they encountered was heavy but inaccurate flak over Taito City, and none of the aircraft was damaged. Colonel Cook painted a generally favorable portrait of the B-32s combat performance in his 1 July summary of the combat test, calling the results of medium and high altitude bombing runs "excellent." Since a major aim of the combat tests was to determine the feasibility of converting all Pacific B-24 units to B-32s, Colonel Cook's report also included a comparison of the offensive capabilities of both aircraft:

The bomb load which can be carried by (the B-32) is roughly two and one-half times that of the B-24. The number of ground and air personnel required per plane is the same for B-32 Groups as for B-24 Groups. This means that, based on effort and hazard to personnel and equipment, the airplane has a 250 percent efficiency advantage over the B-24. All personnel in the theater have recognized this fact. In addition, the airplane is about fifty miles per hour faster and has range capabilities greatly exceeding that of the B-24.

Colonel Cook also pointed out that combat crews were having no difficulties handling the B-32 in the air, and most had found the Dominator to be a rugged and stable bombing platform. And, though Cook's report listed over forty "clean-up" items concerning the layout and suitability of mechanical items, he did not find this number excessive and stated that the maintenance required by the Dominator was "much less than expected."

Lieutenant Colonel Stephen D. McElroy, the representative from Headquarters, USAAF, who flew with Major Britt in 42-108528, was also pleased with the B-32s performance in the combat tests. His summary report stated that "the B-32 airplane in its present condition is suitable for combat operations in this theater" and would be equally suitable for unrestricted combat operations throughout the Far East if minor corrections were made to the Dominator's faulty heating system, nose and tail turret ammunition feed systems. and engine exhaust stacks. Lieutenant Colonel McElroy concluded his report by saying that the B-32's bombing abilities were "excellent," being "much better than expected."

The successful completion of the B-32 combat test led Headquarters, USAAF, to move ahead with plans to convert the 386th entirely to Dominators. Colonel Cook and Lieutenant Colonel Wells were directed to step up the training of ground and air crews,

Strike camera photo of 42-108532's attack on Antatet, 29 May 1945. USAF/AFSHRC

and were informed that a further six B-32 aircraft would be assigned to the 386th in July and August. This would give the unit a total of nine Dominators, making it the first all B-32 squadron in the Army Air Forces. It was anticipated that the 312th Group would move to an advance base on Okinawa as soon as the conversion of the 386th was completed, and would then take part in the final aerial offensive against Japan. In accordance with these directives the personnel of the test contingent and the 386th spent most of July preparing for the Squadron's conversion and the ultimate move to Okinawa. The three Dominators were kept busy with a seemingly endless series of transition flights and practice missions, and the routine was broken by only one combat mission, a July 6th raid on Tako Town, Formosa. All three Dominators took part in the attack, though a "guest bombardier" from Fifth Bomber Command succeeded in putting only six out of thirty-three 1000-pound bombs on target.

The last few weeks of the Second World War were eventful ones for the B-32 combat crews of the 386th Bombardment Squadron. The atomic bombing of Hiroshima on 6 August stirred up rumors of an impending Japanese surrender, but, when no such capitulation was immediately forthcoming, the 386th moved to Okinawa as planned. The advance parties were in place at Yontan Airfield near Sobe by 8 August and, when the atomic attack on Nagasaki the next day also failed to bring about a rapid Japanese surrender, the B-32s at Floridablanca were ordered to Okinawa. "Hobo Queen II" and 42-108528 flew in on the 11th, an engine malfunction having kept "The Lady is Fresh" in the Philippines, and were joined the following day by four of the six new Dominators promised to the 386th by Headquarters, USAAF. These aircraft, 42-108530, 539, 543 and 544, were all B-32-20-CFs, though 42-108578, which arrived at Yontan on 13 August, was a B-32-35-CF. The last of the six, 42-108531, was also a B-32-20-CF, but it was so delayed in transit that it would not take part in any combat missions.

On 13 August Far East Air Forces Headquarters directed the B-32s of the 386th to continue operations against the Japanese despite the defacto cease-fire that had existed in the Pacific theater since the bombing of Nagasaki. On the night of 14/15 August 42-108528 attacked a small vessel during a shipping search of the South China Sea, and a second vessel was sunk during a two-plane sweep later that day. On the evening of the 15th two more Dominators set out to observe enemy air activity over Korea and Honshu, but were recalled when the Japanese agreed to an official cease fire. The B-32s did not stay on the ground for long, however, for FEAF directed the 386th to carry out daylight photo-reconnaissance missions over Tokyo in order to monitor Japanese compliance with the cease fire terms.

The first of these recon missions was flown by "Hobo Queen II" and 42-108543 on 16 August. The flight was not opposed by the Japanese, and the aircraft were able to make their photo runs without interference and return safely to Okinawa. But things did not go quite so well the following day when "Hobo Queen II" led 42-108539, 543 and 578 back over Tokyo. The aircraft, still called

The second edition nose art on "Hobo Queen II," worn from July 1945 through March 1946. Note retracted landing light (below the "N" in Queen) and RCM blade antennas. William T. Y'Blood

"The Lady is Fresh" second edition nose art, which she wore from July through October 1945. Note the retracted landing light directly below the pinup. Lindy LeVine stands at right.

Lindy LeVine via William T. Y'Blood

Top: George Davis (right), flight engineer on 42-108578, proudly points out the three victory symbols added to the aircraft's nose after the August 17/18 Tokyo missions.
Lindy LeVine via William T. Y'Blood

Bottom left: After sustaining fairly heavy damage during the 17 August duel over Tokyo, 42-108539 was relegated to "hanger queen" status and cannibalized for parts. Here she sits rather forlornly beside a Yontan taxi strip, with only one engine and prop still attached.
William T. Y'Blood

Bottom right: The remains of 42-108544 lay in a coral pit at the end of the Yontan runway after the 28 August crash. All thirteen men aboard were killed in the accident.
Lindy LeVine via William T. Y'Blood

"Direct From Tokyo," B-32-20-CF 42-108530, was one of the six additional B-32s assigned to the 386th BS after the conclusion of the 11-mission Dominator combat test. This aircraft did not see any significant combat, but instead was returned to the U.S. in September 1945 as a flying billboard bearing the "Roaring '20s" and Fifth Air Force emblems. The antenna atop the nose was part of the Blind Landing System, while those on the side of the fuselage below the pilot's compartment and in the center of the Fifth's emblem were the receiving and transmitting antennas of the SCR-729 IFF Interrogator-Responder. Bill Cleveland via Dana Bell

Dominators by their crews despite the Army's decision in mid-August to return to the name Terminator, were tracked by Japanese radar during the approach to the capital. Soon after splitting into two-plane elements for the photo run the aircraft were fired upon by flak batteries on the outskirts of the city. The B-32s jammed the radar and the fire soon dropped off, but not before 543 had taken several minor hits in the port wing. A new threat soon appeared in the form of ten fighters. The bombers beat off the attack with only slight damage to themselves, and later claimed one confirmed kill and two probables. "Hobo Queen II" and 42-108578 returned to Tokyo on 18 August, and were again attacked by Japanese aircraft. The American gunners scored two more kills and another two probables, but it was not a one-sided battle. "Hobo Queen II" escaped without damage or casualties, but 42-108578 was badly shot up, her forward upper gunner and one photographer were wounded, and a second photographer, Sergeant Anthony J. Marchione, was killed.

The last Dominator mission of the Second World War was also haunted by tragedy, though not due to the efforts of the Japanese. On 28 August, 42-108544 led three other B-32s (528, 532, and 578) onto the Yontan runway for a special photographic mission to Tokyo. The pilot of 544 began his take-off roll normally, but at the far end of the runway, the plane lost power and the pilot's efforts to brake the heavily loaded aircraft to a halt were fruitless. It skidded off the end of the runway and into a coral pit, killing all thirteen men aboard in the ensuing explosion and fires. The other three aircraft continued the mission despite the loss of 544 and, after completing the recon without Japanese interference, turned back toward Okinawa. Some forty minutes after departing Tokyo 42-108528 lost power on number two engine, and a half-hour later number four also went. The pilot, Second Lieutenant Collins Orton, was able to nurse the rapidly descending bomber to within a few hundred yards of two American destroyers on picket duty in the North China Sea and ordered the crew to bail out. All thirteen men aboard left the stricken bomber just before she exploded; three were picked up by the USS Henley and nine were pulled from the water by the crew of the USS Aulick. One man, Corporal Morris C. Morgan, was never found, and another, Staff Sergeant George A. Murphy, later died of injuries aboard Aulick.

On 30 August 1945 FEAF officially directed the 312th Bombardment Group to cease all B-32 flight operations and begin preparing their Dominators for movement to the United States.

One B-32, 42-108530, was to be flown to New York on a publicity tour. This aircraft, bearing the name "Direct From Tokyo" and large 312th BG and Fifth Air Force insignia on her nose, departed Okinawa on 31 August. The 312th's remaining Dominators left for the U.S. during the first two weeks of October, except for "Hobo Queen II." The venerable aircraft was severely damaged by a nose wheel collapse on 10 October, and was finally broken up on Okinawa in May 1946.

Some returning crews gave their planes special insignia, as the photo on the preceding page shows. "Hobo Queen II" had the Fifth Air Force marking applied to its vertical tail, a list of the flight crew under the pilot's window, and, it seems, some coverup paint

The venerable "Hobo Queen II" at Yontan after her 10 October nose gear collapse. The aircraft was broken up on Okinawa in May 1946.
Lindy LeVine via William T. Y'Blood

for the scantily clad "Queen." Pilot Lyman P. Combs' 42-108543 got a large map of Indiana and the name "Harriet's Chariot," along with the names of the fifteen returning airmen. At least two planes carried large black playing-card figures (clubs) on their tails, the 386th Bomb Squadron's marking.

The B-32's original bombsight mounting was found to be inadequate during the combat test, and was replaced by the modified mount shown here.   USAF/AFSHRC

Detail of lowered B-32 radome, with open nose gear doors at left. The radome housed the AN/APQ-13 Radar Bombing and Search Set (AN/APQ-13A on B-32-30-CF and all subsequent), which was commonly called the H2X. This was a 3cm unit with 29-inch antenna reflectors, and operated in the 10,000 megahertz range to produce a narrow beam in the horizontal plane and a broad beam in the vertical plane.   USAF/AFSHRC

A B-32 of the 386th Bombardment Squadron (Heavy), 312th Bombardment Group, being refueled at Okinawa's Yontan airfield, August 1945. The unit was eventually given nine Dominators. They were B-32-20-CFs 42-108528, 529, 530, 531, 532, 539, 543, 544, and B-32-35-CF 42-108578. The B-32-35-CF differed from the others in having stabilized A-17A nose and tail turrets, Type A-6 portable oxygen bottles, and two built-in bomb bay fuel tanks that raised normal fuel capacity to 6,960 gallons.

The Sperry-built A-17 ball turret supplied the nose and tail defense on the B-32. This electric-hydraulic unit, in which the gunner sat on a small armored seat, behind a thick plate of bullet-resistant glass straddling the receivers of two M-2 .50 caliber machine guns, had angles of fire of 150 degrees in the azimuth plane and 120 degrees in the elevation plane. The turret was attached to the fuselage frame by a bracket at the top and the traversing gear housing at the bottom, the whole ball moving left and right on the pivot points carried by the attaching members. Vertical movement was accomplished by the 2½-foot wide C-shaped center section, which rotated about a lateral axis. This assembly carried the guns, gun sight, armor protection, and all other facilities required by the gunner. Ammunition entered the turret through two ports situated above and to each side of the rear entry hatch, and fed through flexible steel chutes from two ammo boxes mounted in the fuselage aft of the ball.

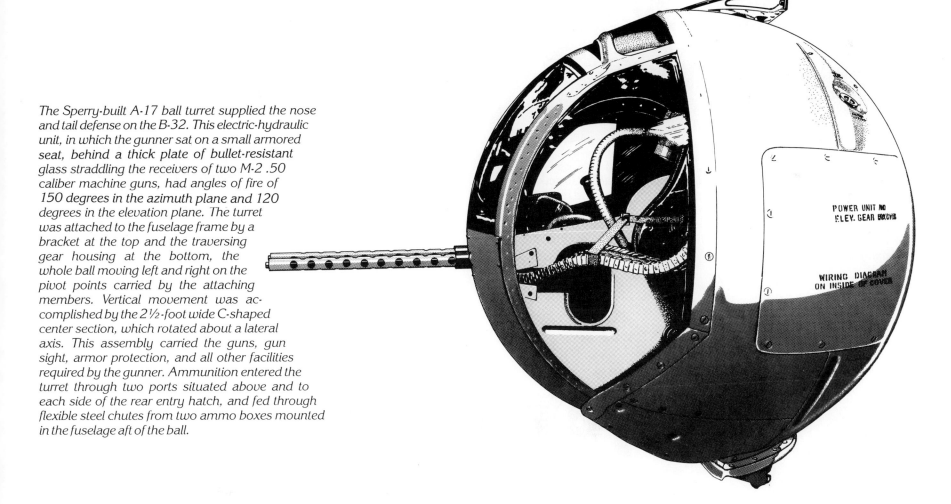

POWER UNIT AND
ELEV. GEAR BOXOVER

WIRING DIAGRAM
ON INSIDE OF COVER

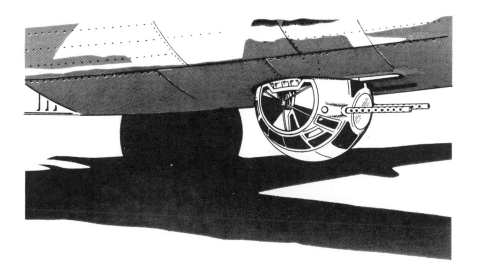

Dominators were equipped with Briggs-built ventral ball turrets. A hand-operated hydraulic system served to extend and retract it. Until scanner windows were added to the sides of the rear fuselage, the lower turret was the only place from which the rear of the engine nacelles and the undersurfaces of the main wing could be clearly seen.

The A-3F dorsal turret, a product of the Glenn L. Martin Co. The guns could be elevated to 79 degrees above horizontal and depressed to a maximum of 6 degrees. The turrets had full 360-degree azimuths, but interrupter gear limited the angles of fire. Teardrop-shaped Plexiglas blisters faired the cupolas when the turrets were in their stowed, fore-and-aft positions. The teardrop and its rainseal/support frame were attached to the turret dome and rotated with it. Beneath the blisters, aft of each turret, soleplate-shaped Alclad structures—which were secured to the airframe and did not rotate—served to finish the streamlining job. The aerodynamic advantage afforded by these bubble assemblies apparently did not outweigh their expense, for military officials recommended that they be eliminated from the design.

This is the second rendition of the nose art on "Hobo Queen II," 42-108532. It was on the left side only, positioned between the two windows of the forward compartment. When she first arrived in the Philippines, the "Queen" had smaller paintings just below the cockpit windows on each side of the fuselage.

CHAPTER SEVEN
Disposal

The end of the war in the Pacific also marked the end of the B-32 Dominator, for the USAAF had no plans to continue the development or use of the troublesome aircraft past V-J Day. The cancellation of the B-32 program had been decided upon by 18 September 1945, and was completed by 12 October with the cessation of production at Consolidated's Fort Worth and San Diego plants. The last six B-32 aircraft (42-108579 through 584) were declared excess upon delivery in September and October, and were flown directly to Reconstruction Finance Corporation (RFC) disposal centers. An additional twelve aircraft in shop-assembled status at Fort Worth and San Diego were declared Terminal Inventory and also flown directly to RFC disposal centers. At least 50 other aircraft that had been partially assembled at the two plants were dismantled by the contractor and sent to disposal sites for scrapping. Those Dominators already accepted by operational units were, if flyable, sent to the closest RFC center, while non-flyables were disposed of in place.

The majority of B-32 aircraft sent to RFC centers were disposed of by 1947. The U.S. government did not consider selling the surplus Dominators to friendly foreign powers, and the B-32's complexity and somewhat exaggerated reputation for mechanical unreliability kept civilian interest to a minimum. Indeed, there is only one recorded case of a civilian attempting to purchase a surplus Dominator for private use. In July 1947 Milton J. Reynolds, the pen manufacturer and adventurer, announced that he had arranged with the War Assets Administration for the purchase of a single surplus B-32. Reynolds intended to use the aircraft for a proposed flight around the world over both poles, but the expedition was eventually cancelled and the aircraft was never delivered.

The Dominator fared no better as a museum piece, for not a single example of the aircraft was preserved for posterity. The one Dominator intended for display at the Air Force Museum, B-32-1-CF 42-108474, was unaccountably declared excess and destroyed at Davis-Monthan Air Force Base in August 1949. In late 1983, the only known remaining B-32 artifacts are a nose turret belonging to the National Air and Space Museum, a partial instrument panel held by a private collector, and a single Alclad wing panel used as a monument to aviation pioneer John J. Montgomery in San Diego.

These Dominators, some of the 50 planes undergoing final assembly at Consolidated's Fort Worth plant when the AAF cancelled the B-32 contracts at the end of the war, were on the average 88 percent complete. But, rather than completing them to fly-away status, the contractor, under AAF direction, dismantled them, loaded them on freight cars, and sent them to the Metals Reserve Company, Camp Howze, Gainesville, Texas, for scrapping. The 50 planes bore serial numbers 42-108595 through 644 and were to have been B-32-40, -45, -50, and -55-CF models.　　General Dynamics